Flowmeters in Water Supply

AWWA MANUAL M33

First Edition

FOUNDED 1881

American Water Works Association

American Water Works Association
6666 West Quincy Ave.
Denver, CO 80235

Printed in USA

ISBN 0-89867-473-5

Contents

Foreword

This manual was prepared as an introduction to the most common flowmeters used today in water supply systems. The flowmeters discussed in this manual include the Venturi, modified Venturis, orifice plate, magnetic, turbine and propeller, sonic, vortex, and averaging Pitot. The discussion of these meters covers basic theory, installation, maintenance, and advantages and disadvantages. General concepts applicable to flowmeters are also discussed, including flow characteristics, precautions on installation, flow straighteners, associated wiring, power and electronics, and data logging.

The manual can be used as a bridge to other literature on flowmeters, to prepare the reader for further investigations into instrumentation design and applications. The cited references will provide excellent information sources. For additional information, the reader should acquire the relevant manuals from the meter manufacturers. These manuals contain comprehensive information on meter specifications, theory, sizing, handling, installation, power and wiring, operation, maintenance, troubleshooting, and parts.

While this manual attempts to include recommended practice in the use of flowmeters, it is not intended as an AWWA standard.

* * *

This manual was prepared by the Automation and Instrumentation Committee of the AWWA Engineering and Construction Division. The members of that committee at the time this manual was approved include

R.W. Graupmann (Chairman), EMA, Inc., Tucson, Ariz.
G.L. Craft (Staff Liaison), American Water Works Association, Denver, Colo.
J.R. Daneker, BIF Industries, Inc., West Warwick, R.I.
J.L. Doane, Portland Water Bureau, Portland, Ore.
R.V. Frykman, Metcalf & Eddy, Inc., Arlington Heights, Ill.
J.W. Garrett, Westin Engineering, Inc., San Jose, Calif.
H.D. Gilman, Greeley and Hansen, Philadelphia, Pa.
O.E. Hopkins (Board Liaison), Salt Lake City, Utah
J.A. Lefever, F.B. Leopold Company, Inc., Leola, Pa.
E.F. Morey, The Turbitrol Company, Atlanta, Ga.
M.J. Okey, CH2M Hill, Inc., Denver, Colo.
R.K. Weir, Denver Water Department, Denver, Colo.

Chapter 1

Characteristics of Flow

Flow is the most important measurement made at water supply facilities, including treatment plants and distribution systems, and flowmeters are used to measure this flow. Flowmeters measure product output of a plant, water distribution into a community, pacing of chemical feeds, and customer charges, to name just a few applications. Day-to-day operational decisions and long-term planning are based on the measurements from flowmeters.

FLOW VELOCITY

The physical properties of fluids important to flowmetering are

- pressure—force per unit area,
- weight density—weight per unit volume,
- viscosity—fluid's resistance to flow, and
- velocity—fluid's speed of flow.

The most important of these is velocity. Of interest to the operator is the average velocity of a cross section of the stream—a thin disk of fluid moving along the pipe axis.

Measuring Velocity

A transducer is needed to detect liquid movement, that is, the velocity of a liquid. A transducer is a device that, by its inherent nature, changes when it is in the path of or is influenced by flow. The transducer, in turn, allows detection of a change in flow. An example is the turbine meter, in which the flow impinging on the turbine vanes imparts energy to the turbine and causes it to rotate. The rotation is proportional to the flow velocity, enabling the user to arrive at a value of velocity by counting the rotations. Each rotation represents a defined volume of liquid passing the meter. Flow is determined by knowing the rate of propeller rotation with time.

Equation of volume flow. In engineering terms, flow is simply the time rate of change of volume of a moving fluid at a pipe cross section. To achieve a time rate of change of volume Q in most meters, the average fluid velocity V and the

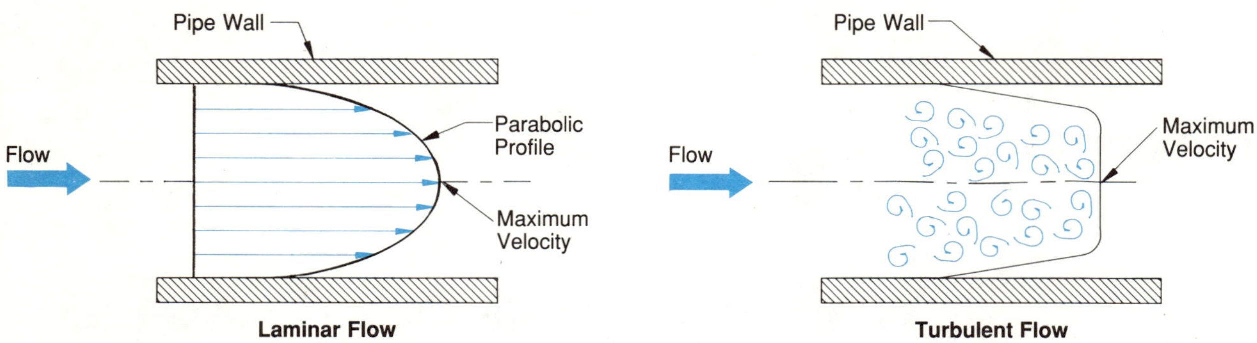

Figure 1-1 Velocity profiles.

cross-sectional area *A* must be known. And most meters are identified according to the physical parameter used to determine the velocity.

The flowmeters considered in this manual are for closed-conduit, full-flowing pipes, thereby allowing the flow cross-sectional area to be that of the pipe or meter. The product of velocity and cross-sectional area that yields volume flow, $Q = VA$, is the basic equation of flow used to yield dimensions such as gallons per minute, cubic metres per second, and million gallons per day.

Laminar and Turbulent Flow

As shown in Figure 1-1, pipe flow can assume two profiles, laminar flow and turbulent flow, with a transition phase between the two. Laminar flow results from slow average velocities. The fluid moves in concentric, cylindrical layers, with the fastest-moving layers toward the central axis and the slowest-moving layers toward the pipe wall. A parabolic profile of flow exists across any diameter of the pipe.

Turbulent flow results from high average velocities. The flow layers disappear, and the velocity across the stream is more uniform. The velocity profile is flattened toward the pipe's central axis. Although the flow as a whole and an average moves in a straight line down the pipe, individual fluid particles zigzag and gyrate.

In the transition phase between laminar flow and turbulent flow, the change is gradual with no distinct transition point.

In the water industry, flows are generally turbulent, and the assumption of turbulent flow applies to the meters described in this manual.

REYNOLDS NUMBER

Closely associated with flow properties is the Reynolds number. The Reynolds number is the ratio of inertial forces, such as from a pump or gravity, to viscous forces, the internal fluid resistance to flow, in a flowing liquid. As a ratio of force to force, the parameter is dimensionless. A Reynolds number provides an indication of which forces dominate the flowing liquid.

Type of Flow

The type of flow, laminar or turbulent, is related to the Reynolds number. Low values (Reynolds numbers less than 2000) indicate that the viscous forces are dominating the flow. Adjacent fluid particles move along in essentially parallel paths. Flow with

Reynolds numbers between 2000 and 4000 are in the transition phase. The velocity profile may fluctuate between laminar and turbulent. At Reynolds number values above 4000 (not a strict borderline value) the inertial forces begin to dominate; the flow profile becomes blunter; the particles all begin moving at the same velocity, except at the walls. Adjacent particles are in more or less random motion, not moving in parallel paths in the turbulent flow. Reynolds numbers will be noted in textbooks and vendor literature as having some impact on meter accuracy and the meter coefficient of discharge. The meters in water supply are designed for use at high Reynolds numbers.

METER COEFFICIENT OF DISCHARGE

Manufacturers establish a generic meter coefficient of discharge *C* for each type of meter produced. This coefficient is directly related to the accuracies found in the manufacturers' published literature.

The coefficient of discharge *C* is used to account for deviations from the true average velocity and the cross-sectional area of the pipe. The ratio of actual flow rate (determined by high-accuracy measurement) to the theoretical flow (calculated from a mathematical model published by the manufacturer) defines the meter's discharge coefficient. A discharge coefficient of unity (1) would indicate a perfect flowmeter. This has never been achieved. However, a review of the literature demonstrates that the closer the coefficient is to 1, the more precisely the meter is constructed and the more accurate the device tends to be under varying operating conditions.

A precise flow calculation is frequently dependent on much more than the obvious variables entered into the mathematical model. If any flow parameters change, the flow calculation may no longer be valid. Significant changes in fluid temperature, density, and specific gravity will cause errors in flow calculations.

No attempt is made to detail the formulas for the mathematical models of specific flowmeters. These are available in the textbooks referenced (see bibliography at the end of this manual).

Chapter 2

Signal Output and Transport

The instrumentation found throughout a water treatment facility provides information to the facility's data acquisition and control system. Instrumentation, or primary sensors, could be the yardstick in a tank, indicators on a complex control panel, or a video-displayed graphic diagram. No matter what the level of sophistication, primary sensors are the critical link to process and equipment operations. In water treatment and distribution, flow is the most important variable, and flowmeters are the most important primary sensors.

FLOWMETER OUTPUT

A flowmeter indicates a physical or electrical change that is proportional to flow. If the flowmeter reading is sent to a modern analog panel and digital computer system, the initial physical response must be converted to a signal that has sufficient energy to drive related equipment. This equipment may be an indicator, controller, recorder, or computer input.

Because there are so many applications for any sensor output, the instrumentation and control industry has defined a standard—the 4–20 mA DC current loop. In a properly designed loop, the power source and the sensor are connected in a series electrical circuit to the items using the signal. As an example, if these items require a voltage input, where the standard is 1–5 V DC, a 250-ohm resistor in the series loop and across the input to the computer or recorder will provide the proper signal.

Signal Conditioning

The raw initial signals from flowmeters are subjected to signal conditioning, which ultimately generates the standard 4–20 mA DC signal and also performs many other functions. These may include filtering, amplification or attenuation, and linearization. A common signal-conditioning operation for differential-pressure flowmeters is to convert the raw signal by a square root function to one directly proportional to flow velocity. It is not uncommon to multiply the velocity and pipe cross-sectional area as part of signal conditioning to generate the desired volume flow.

Signal enhancement. Microprocessor technology and subminiature electronics are being applied to the conditioning of the signal to enhance the application of flowmeters through programmed functions and menu readouts. Some of these enhancements include

- digital rate display,
- digital totalizer display,
- selection of flow spans,
- selection of engineering units in digital display,
- filtering of signal,
- remote location display,
- auto calibration,
- auto diagnostics with alarm readouts, and
- coded error messages.

In some units a keypad is included to enable the user to select the desired functions. These devices are readily available as small low-power, low-cost electronic packages.

TELEMETRY

Through the technology of the current loop, the meter output can be read not only locally but at significant distances from the source. The limitation of the current loop is the total resistance in the circuit, including the 250-ohm resistors across inputs to equipment, and the wire length.

Where distances or conditions reduce the practicality of direct wire loops, the techniques of telemetry are used. Thus, measured physical quantities can be transmitted to remote locations, in a form suitable for display and analysis. The same 4–20 mA DC loop can include a device to make the signal acceptable for remote transmission with virtually no reduction in accuracy. The conversion may be to transmission by FM radio, microwave, fiber optics, telephone lines, or other forms of telemetry media.

Applications

Common applications of telemetry today are the monitoring and control of remote pump station flows from a central location. Telemetry brings the flows and related signals into the control panel for evaluation and response, either manually or by computer logic. As needed, a control response is transmitted via telemetry back to the remote pump station.

The capabilities of telemetry encompass far more than remote monitoring of flows. Through telemetry and computer installations, centralized monitoring and control of entire water supply, treatment, and distribution systems are feasible and in operation today.

Chapter 3

Installation Precautions

The reliability and accuracy of a flowmeter's output signal are highly dependent on meter installation. An improperly located and installed meter will degrade the inherent specified accuracy below an acceptable level. Responsible manufacturers provide guidance and recommendations for meter installation, which should be rigorously followed. Nevertheless, the meter must be checked immediately after installation and periodically thereafter to assure the stability of calibration and accuracy.

PIPING CONFIGURATIONS

Approach piping and, in some cases, discharge piping configurations are important and often neglected installation considerations. All meters have limitations due to piping configurations, and too often this fact is ignored in piping layouts.

To prevent degeneration of accuracy due to piping configurations, manufacturers recommend straight pipe lengths both upstream and downstream. The user should acquire manufacturer's tables, which list minimum length of straight pipe required for installation on both sides of the meter.

If manufacturer's specifications are unavailable, a length of 50 diameters of straight pipe upstream from a meter, although very conservative, is a safe distance for installation of flowmeters discussed in this manual.*

Fittings

Meters should not be placed in close proximity to a bend, valve, or other fitting that is likely to disturb the flow condition at the meter. This is because the calibration usually takes place with the meter mounted downstream of a long length of straight pipe. If this recommendation is not followed in field installation, the manufacturer's meter calibration may be invalidated. Downstream of the meter, the presence of

*This information according to Hayward, A.T.J. *Flowmeters, A Basic Guide and Source Book for Users*. John Wiley & Sons, New York (1979).

bends, valves, and fittings may also cause disturbances, which can be transferred back to the meter and also affect meter accuracy.

FLOW STRAIGHTENERS

Where installation of the recommended straight pipe length is impossible, a straightener to remove flow distortions, in particular, swirl, should be installed. Swirl, caused by elbow bends or certain pumps in the piping, is a rotary motion of the flow superimposed on the forward motion, which adversely impacts accuracy of most flowmeters.

Flow straighteners include the tube bundle, perforated plate, étoile, and the air moving and conditioning association standard (AMCA), as shown in Figure 3-1.

Installation

To avoid causing more harm than good, flow straighteners must be properly installed in the proximity of both the source of flow disturbance and the flowmeter itself. In general, straighteners should be inserted at least three pipe diameters downstream of the disturbance source, where some dissipation of the disturbance will have occurred. The distance between the straightener and the meter should ideally be 20 pipe diameters, but not less than 10 diameters, to allow most of the distortion to disappear. In general, the AMCA and étoile flow straighteners produce only negligible head loss. The tube bundle and perforated plate produce substantial head loss.

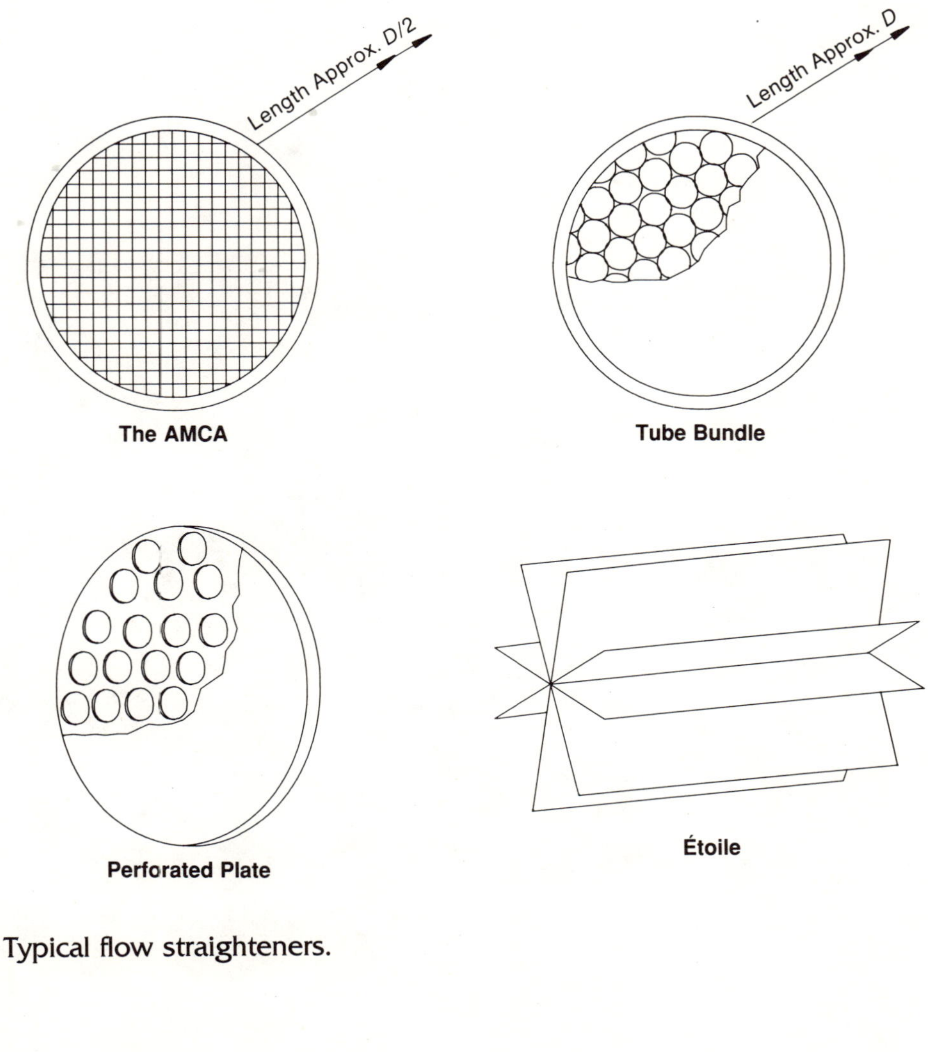

Figure 3-1 Typical flow straighteners.

ELECTRICAL CONNECTIONS AND ELECTRONICS

Today virtually all flowmeter installations generate electrical signals to represent the flow measurement, typically the 4–20 mA standard analog of the specified flow range. Consequently, special attention must be given to electrical connections and secondary instruments during installation. In particular, pressure and temperature ratings must not be exceeded. Equipment must be protected from corrosion. If the meter is to be installed below ground level, it must be capable of withstanding total immersion, with protection provided for associated electrical appurtenances.

Precautions

The following precautions should be taken during meter installation:

- provide preamplification as part of conditioning the signal generated by the sensor. This is done in order to create a high signal-to-noise ratio and, thereby, make the meter output signal less susceptible to noise distortion;
- use shielded, twisted pair wiring to transmit the signal, with the shield grounded only at the signal transmitter end to prevent ground loops;
- maintain a large distance between the signal wire and power cable, and allow crossings, where necessary, only at right angles; and
- maintain an inspection vigil for poor connections, damaged or worn components, and the effects of corrosion.

Chapter 4

Types of Flowmeters

The most common flowmeters used in today's water treatment facilities are discussed in this chapter. These include the Venturi and modified Venturis, orifice plate, magnetic, turbine and propeller, sonic, vortex, and averaging Pitot flowmeters.

VENTURI FLOWMETERS

Accuracy*: ±0.75 percent of rate
Repeatability†: ±0.25 percent
Rangeability‡: 4:1 and up to 10:1
Size range: 1–120 in. (25 mm–3 m)
Head loss: Low
Relative cost: Medium

The most commonly used flow-measurement devices are the differential-pressure flowmeters, including the Venturi meter, or Venturi. Their popularity is largely due to a combination of flexibility, simplicity, ease of installation, and reliability.

In the Venturi, a defined constriction (throat) within the meter body causes an increase in flow velocity at the constriction, resulting in a corresponding decrease in pressure at that point (Figure 4-1). The difference in pressure between the upstream pressure connection (piezometer) and the pressure connection at the constriction (throat) is proportional to the square of the flow. Flow is then calculated from the square root of the measured pressure differential.

**Accuracy:* The degree to which an indicated value matches the actual value of a measured variable. In process instrumentation accuracy is the degree of conformity of an indicated value to a recognized accepted standard value, or ideal value.

†*Repeatability:* The closeness of agreement among a number of consecutive measurements of the output for the same value of the input under the same operating conditions, approaching from the same direction, for full-range traverses.

‡*Rangeability:* The ratio of the maximum flow rate to the minimum flow rate of a meter.

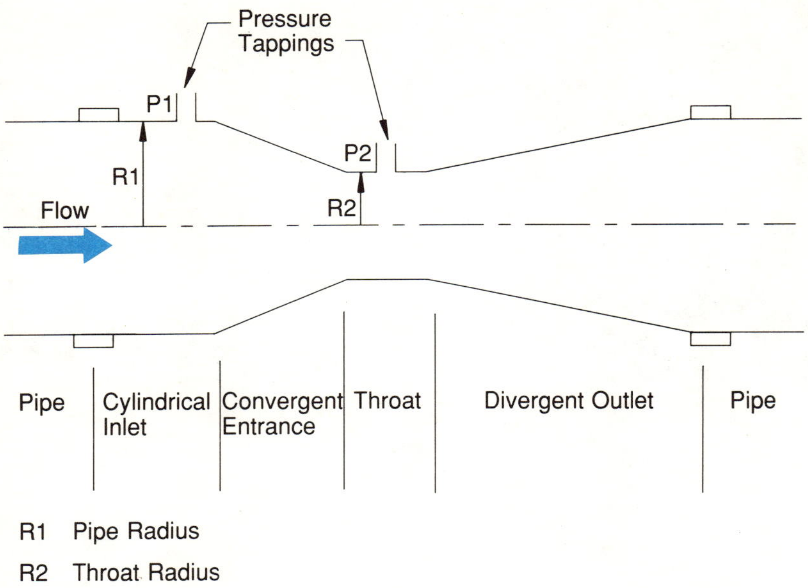

Figure 4-1 Schematic of a Venturi meter.

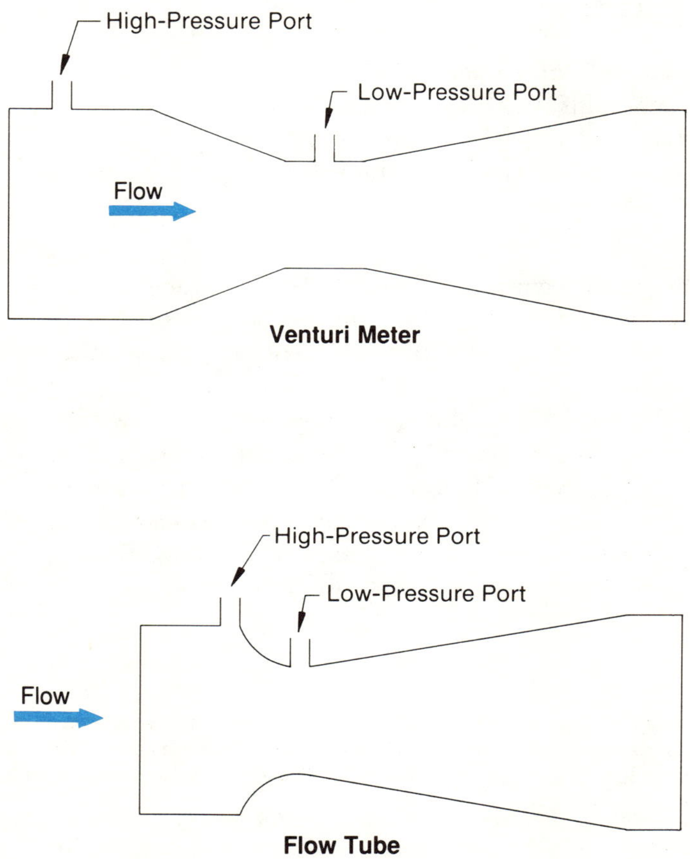

Figure 4-2 Location of pressure ports in Venturi meter and flow tube.

To obtain an electrical signal for flow measurement, the pressure at the ports must be transferred to a differential-pressure transducer. The transducer is typically a flexible diaphragm with a small chamber; one of the two pressure lines from the meter tube is connected to each side. The pressures each exert a force on the diaphragm and cause deflection of the sides. To sense the deflection of the diaphragm, a strain gauge, or a variable capacitance or a variable reluctance device is built into the transducer. This device generates an electrical signal proportional to the pressure difference across the diaphragm. Direct-reading differential pressure may be detected and scaled using a U-tube manometer or double-bellows gauge.

It is important to note that the major distinction between the Venturi and modified forms of this device, such as flow tubes and flow nozzles, is that the pressure connections are pure static measurements taken at points where the direction of fluid flow is not changing (see Figure 4-2). The shape of the approach to the throat conditions the flow pattern in the throat. With this hydraulic profile conditioning built into the body inlet structure, only the shortest lengths of straight pipe on the inlet side are necessary, when compared to other primary flowmeters. For the same pressure differential and the same size pipe and throat constriction, the Venturi yields 60 percent less head loss than the orifice plate meter.

The accuracy of a Venturi is ±0.75 percent of actual flow. This is exclusive of any additional tolerances introduced by instrumentation used with the meter to measure the differential pressure. Contrary to much of the published data, the Venturi has no range limitations up to the speed of sound. However, the instrumentation does have definite limits. Some instruments can provide a 10:1 range, several 8:1, and many only 4:1.

The Venturi meter has a low permanent head loss and does not obstruct the flow of suspended matter. The coefficient of discharge of the classic Venturi is 0.983, which remains constant regardless of pipe or throat size. The Venturi has the most comprehensive documentation regarding its coefficient, with hundreds of thousands of laboratory test points taken in every conceivable piping configuration.

The Venturi requires the greatest laying length of all differential-pressure meters and is, therefore, the heaviest flow tube as well. Venturis are commonly constructed from cast iron and naval bronze.

Installation

Before installing a Venturi, there are factors to consider in the selection process.

- Select a meter with the highest differential pressure, while considering how much head loss can be tolerated. A high differential provides the greatest energy for driving the instrumentation, which improves the range and accuracy.
- Select a meter with the smallest throat size in any given line size, again considering head loss. In the literature and manufacturer's tables, throat size is defined in terms of the beta ratio—the ratio of throat diameter to pipe diameter.
- Review the upstream piping configuration. A Venturi is not affected by the downstream configuration except for a slight increase in head loss.
- Consider future expansion of the facility, which may increase required flow.

Installation of a Venturi is critical to the accuracy of the differential-pressure measurement. Errors in computing flow from the differential pressure may become unacceptable if the piping arrangements create distorted flow. Swirls or vortices that affect meter accuracy can be produced by a projecting gasket, misalignment, or a burr on a pressure tap.

Preferably, the Venturi should be installed with its axis horizontal, and the fluid should enter the tube with a fully developed velocity profile that is free from swirls and vortices. In a horizontal installation, the pressure port tappings must not be at the bottom, subject to clogging, nor at the top, subject to air bubble trapping. The preferred location is on the side in the horizontal plane of the centerline.

Maintenance

As a flowmeter with no moving parts, the Venturi would normally require less attention than, for example, a turbine meter. However, there can be extensive piping, fittings, and valves in the pressure-differential assembly. Lines may clog and corrosion may begin to appear. Periodic disassembly, inspection, and cleaning should be practiced. The pressure sensors, in particular, should be removed and inspected. While it is especially important to inspect the throat of the Venturi for debris or deposits, the high fluid velocity usually scours the throat.

Manufacturers provide step-by-step instructions for checking out the meter components. The procedures include disassembly, inspection and testing, parts replacement, and reassembly. The emphasis is on the differential-pressure unit and electrical housing. Instructions for meter zero and span adjustment are also included. A troubleshooting guide may be provided with symptoms, potential sources, and recommended corrective action. To aid the user, illustrated drawings, schematic diagrams, and parts lists are provided in the manufacturer's manual. Figure 4-3 provides an example of a troubleshooting guide for a differential-pressure transducer.

Advantages and Disadvantages

The following are advantages of the Venturi meter:

- life expectancy of a Venturi with manometer is documented to be greater than 50 years (this may not be true of modern instrumentation in general, although some earlier instruments were good for 30 years or more); materials of construction are well documented for long life;
- simplicity of construction, including no moving parts;
- no sudden change in contour and no sharp corners;
- relatively large pressure recovery in the recovery cone, resulting in low head loss and leading to substantial power savings where large flows are metered; and
- heavily documented in the literature as an acceptable type of flowmeter.

The following are disadvantages of the Venturi tube:

- larger units are costly to purchase and install;
- largest and heaviest of the differential-pressure meters; and
- differential pressure is not linear with flow rate, requires square root extraction, and reduces rangeability.

MODIFIED VENTURIS

Several flowmeters are derived from the Venturi and operate on the same principle. They are classified as differential-pressure producers, and include flow tubes, or "low loss" (proprietary) flow tubes. The object of any modification to the Venturi is to achieve shorter laying length, less cost, and/or higher head recovery. The length of the throat in the flow tubes is much shorter than in Venturi meters. This may result in a less stable flow pattern through the throat, which will affect accuracy.

Troubleshooting—Differential-Pressure Transducer

SYMPTOM: High Output

POTENTIAL SOURCE AND CORRECTIVE ACTION

1. **Primary Element**
   Check for restrictions at primary element.
2. **Impulse Piping**
   Check for leaks or blockage.
   Check that blocking valves are fully open.
   Check for entrapped gas in liquid lines and for liquid in dry lines.
   Check that density of fluid in impulse lines is unchanged.
   Check for sediment in transmitter process flanges.
3. **Transmitter Electronics Connections**
   Make sure bayonet connectors are clean and check the sensor connections.
   Check that bayonet pin #8 is properly grounded to the case.
4. **Transmitter Electronics Failure**
   Determine faulty circuit board by trying spare boards.
   Replace faulty board.
5. **Sensing Element**
   See Sensing Element Checkout Section.
6. **Power Supply**
   Check output of power supply.

SYMPTOM: Erratic Output

POTENTIAL SOURCE AND CORRECTIVE ACTION

1. **Loop Wiring**
   Check for adequate voltage to the transmitter.
   Check for intermittent shorts, open circuits, and multiple grounds.
   NOTE: Do not use over 100 V to check the loop.
2. **Process Fluid Pulsation**
   Adjust electronic damping pot (4–20 mA DC only).
3. **Impulse Piping**
   Check for entrapped gas in liquid lines and for liquid in dry lines.
4. **Transmitter Electronics Connections**
   Check for intermittent shorts or open circuits.
   Make sure that bayonet connectors are clean and check the sensor connections.
   Check that bayonet pin #8 is properly grounded to the case.
5. **Transmitter Electronics Failure**
   Determine faulty board by trying spare boards.
   Replace faulty circuit board.

SYMPTOM: Low Output or No Output

POTENTIAL SOURCE AND CORRECTIVE ACTION

1. **Primary Element**
   Check installation and condition of element.
   Note any changes in process fluid properties that may affect output.
2. **Loop Wiring**
   Check for adequate voltage to transmitter.
   Check for shorts and multiple grounds.
   Check polarity of connections.
   Check loop impedance.
   NOTE: Do not use over 100 V to check the loop.
3. **Impulse Piping**
   Check that pressure connection is correct.
   Check for leaks or blockage.
   Check for entrapped gas in liquid lines.
   Check for sediment in transmitter process flange.
   Check that blocking valves are fully open and that bypass valves are tightly closed.
   Check that density of fluid in impulse piping is unchanged.

Adapted from Rosemount Inc., Eden Prairie, Minn.

Figure 4-3 Excerpt from differential-pressure transducer troubleshooting procedures.

Although similar to the Venturi, the flow tube has a relatively short transition section from the inlet to the throat (see Figure 4-2), and a recovery cone of gradually increasing diameter. The flow tube causes significantly less pressure drop than the orifice plate flowmeter (discussed later), but it is costlier. With its shortened inlet section, the flow tube is usually less expensive than the Venturi for the same line size. The body structure does not provide hydraulic profile conditioning, and the flow tubes are more sensitive to upstream flow disturbances.

A very low head loss, lower than that for a comparable Venturi with the same beta ratio, has been achieved by the "low-loss" flow tube. The head loss is a percentage of the differential, based on throat size.

Because the materials of construction for the flow tube are the same as those of the Venturi, long life may be anticipated. Some flow tubes will lose their discharge coefficient at relatively low and high Reynolds numbers, but only the former is important in the water market. The accuracy is ±1 percent of the flow rate, and the range, like the Venturi, is limited only by the associated instruments.

Another variation on the Venturi meter is the insert flow tube. The convenience of the unit is based on its short laying length, single mounting flange with no inlet cone upstream of the flange, and low head loss. Both the high- and the low-pressure metering taps are built into the flange.

ORIFICE PLATE FLOWMETERS

Accuracy: ±1/4 to ±2 percent of full scale
Repeatability: ±0.25 percent
Rangeability: 4:1
Size range: All
Head loss: Medium
Relative cost: Low

The orifice plate flowmeter consists of a thin plate with a hole in it (Figure 4-4). The standard plate is a circular disc, usually of stainless steel, 1/8–1/2 in. (3–13 mm) thick, and containing an orifice with a sharp leading edge. The hole is usually concentric with the pipe into which the plate is inserted and perpendicular to the axis

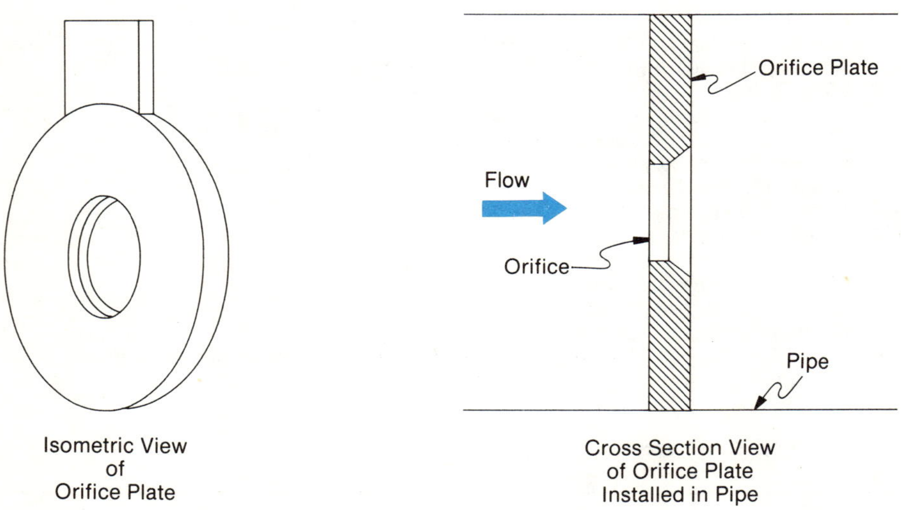

Figure 4-4 Schematic of an orifice plate flowmeter.

of flow. The plate is installed in a pipeline between two flanges, and the differential pressure across the plate is measured. Because of its simplicity, low cost, ease of installation, and reasonable accuracy, the orifice plate is among the most common primary elements for metering flow.

Most orifice plate meters for clear water are made with a circular orifice concentric with the pipe. In special applications, notably for flowing solids, the orifices may also be eccentric or segmental. Passage of entrained solids is permitted if the hole is tangent to the inside surface of the bottom of a horizontally laid pipe.

As in a typical differential-pressure flowmeter, the pressure upstream (approximately one pipe diameter from the plate) is compared with the pressure downstream at the point of narrowest streamflow. The narrowest section of flow is called the *vena contracta*, and is taken to be one half pipe diameter downstream from the plate. These points define where to locate the pressure taps.

Numerous eddies form downstream from the plate between the pipe wall and the *vena contracta*, causing kinetic energy to be dissipated as heat. This accounts for the medium to high head loss associated with the orifice plate meter.

The orifice plate size can be fabricated to accommodate any pipe size. Inherent accuracy, independent of pipe diameter errors and the differential-pressure sensor, may be $\pm 1/4$ percent to ± 2 percent of full scale.

Orifice plates are also available as part of integrated pipe assemblies, which include the plate, pipe length, pressure taps, and, in some cases, the differential-pressure sensor and signal transmitter.

Installation

The orifice plate is usually mounted between a pair of flanges. Manufacturers may extend the plate to include a tab above the edge of the pipe flange. The tab, suitable for a nameplate, may contain pertinent data on the specific installation and may identify the upstream side. To prevent errors in flow measurement, the gaskets should not protrude across the plate face beyond the inside pipe wall. Typically, the orifice plate also requires a straight run of smooth flow before and after the plate. Pressure taps must be installed perpendicular to the pipe wall. For horizontal pipe runs, the pressure taps should be in the horizontal plane of the pipe centerline. Burrs and intrusions at the tap must be removed.

Maintenance

As a differential-pressure flowmeter with no moving parts, the orifice plate meter requires maintenance similar to that for the Venturi. The orifice should be visually checked periodically to be sure its dimensions and sharp leading edge are unaffected by the flow. Degeneration of the sharp edge can result in significant errors in the measurement. Solids may collect behind the plate at the bottom and gases may be trapped at the top, which can be cleared by removal of the plate. Some manufacturers provide special mounting devices to allow the orifice plate to be inserted and removed from the pipe without interrupting the flow, which simplifies maintenance.

The pressure taps should also be examined for possible obstruction. The pressure-differential assembly should be periodically checked.

Advantages and Disadvantages

The following are advantages of orifice plate flowmeters:

- lowest cost differential-pressure meter;
- economically manufactured to very close tolerances;

- easy to install and replace; and
- no moving parts.

The following are disadvantages of orifice plate flowmeters:

- high permanent pressure loss through orifice plate can cause significant power costs;
- volume flow is nonlinear;
- rangeability is lower than with linear-output flowmeters; and
- long upstream and downstream straight pipe runs are required.

MAGNETIC FLOWMETERS

Accuracy: ±0.5 percent of rate
Repeatability: ±0.25 percent
Rangeability: 10:1
Size range: 0.1–120 in. (2.5 mm–3 m)
Head loss: None
Relative cost: High

In a magnetic flowmeter (Figure 4-5), a magnetic field is generated around an insulated section of pipe. Water passing through the magnetic field induces a small electric current that is proportional to the velocity of the water flow. The electric current is measured and converted to a numerical indication of water flow. The amount of the voltage produced is approximately 0.05 V/ft/s for the velocity of the fluid, and the output is essentially linear. The liquid moving through the meter must have sufficient ion content to provide a conductivity of 20 μS/cm. Water treatment plant flows meet this conductivity requirement.

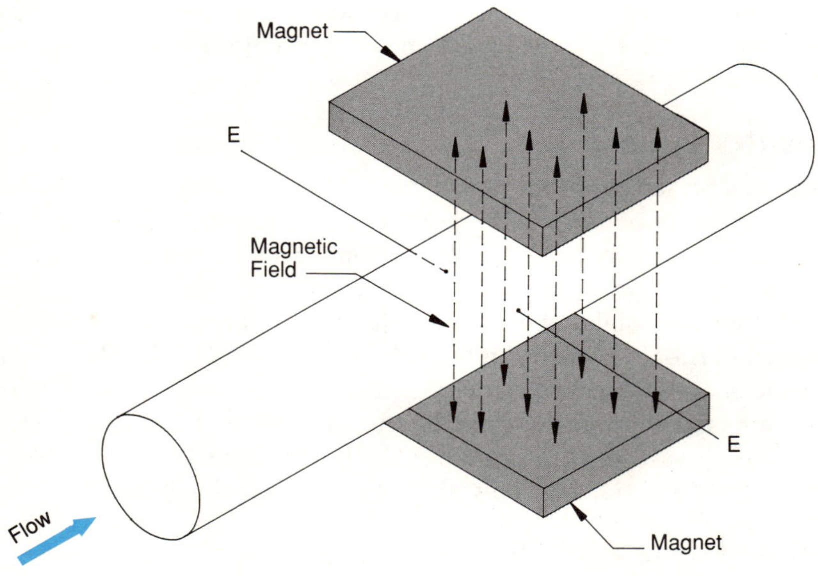

Figure 4-5 Schematic of a magnetic flowmeter.

The meter's accuracy is ±0.5 percent of the actual flow rate; it can be used over a velocity range of 0.2–35 ft/s (0.06–11 m/s). The range of a magnetic flowmeter is 10:1. For clean water, the allowable velocity range is from 1 ft/s (0.3 m/s) minimum to 10 ft/s (3 m/s) maximum.

The head loss across a magnetic flowmeter is the loss that is caused by the length of pipe forming the body of the meter. Head loss would be increased if it were necessary to reduce pipe diameter and incorporate a smaller size meter in order to raise the flow velocity in the meter to an acceptable level.

A magnetic meter has no constriction in its cross-sectional area, so the flow profile is not affected. Hence, whatever profile error may exist in the piping is carried through the metering area. However, area averaging makes the magnetic flowmeter less sensitive than other types of flowmeters to profile changes. The magnetic meter does not have a discharge coefficient definable by its hydraulic shape. Consequently, all magnetic meters should be calibrated, even those of the same line size.

Sizes of magnetic flowmeters range from 0.1 to 120 in. (2.5 mm to 3 m) in diameter. Following are some of the obvious advantages for this type of meter. The magnetic meter has a short laying length, and is much lighter in weight than a Venturi meter. The meter is constructed of stainless steel; therefore, it has a long life. In addition, the solid-state electronics that generate the flow signal should also have a long life.

Installation

A magnetic flowmeter should be installed with the electrodes located at the ends of a horizontal diameter rather than a vertical diameter. This will ensure continued electrode immersion even when air bubbles are present in the water flow.

The following installation steps should be taken in order for the meter to function properly:

- select a meter that has a reasonable velocity at maximum flow (on the order of 5 ft/s [1.5 m/s] and a velocity at the minimum flow not less than 1 ft/s [3 m/s]);
- provide suitable upstream piping length, as recommended by the meter manufacturer;
- ensure proper grounding isolation between the meter body and the pipeline, to avoid transient voltage interference;
- provide 120 V AC power input at about 100 W (models are available for 220 V); and
- use electrical bonding strips to bypass the currents around the meter, if the meter is installed in a pipe that is part of an electrogalvanic-corrosion prevention system.

Maintenance

Manufacturers' troubleshooting and maintenance procedures for magnetic flowmeters are frequently documented on logic flowcharts. These charts include functional statement blocks and "Yes–No" arrows that indicate the next step in the checking sequence. By using the chart, the meter repair person is directed to the problem sources, diagnostics, and recommendations for remedial action. Flowcharts may be provided for meter installation, and primary, converter, and power supply checks. Figure 4-6 (page 18) is an example of a troubleshooting flowchart used to check meter installation.

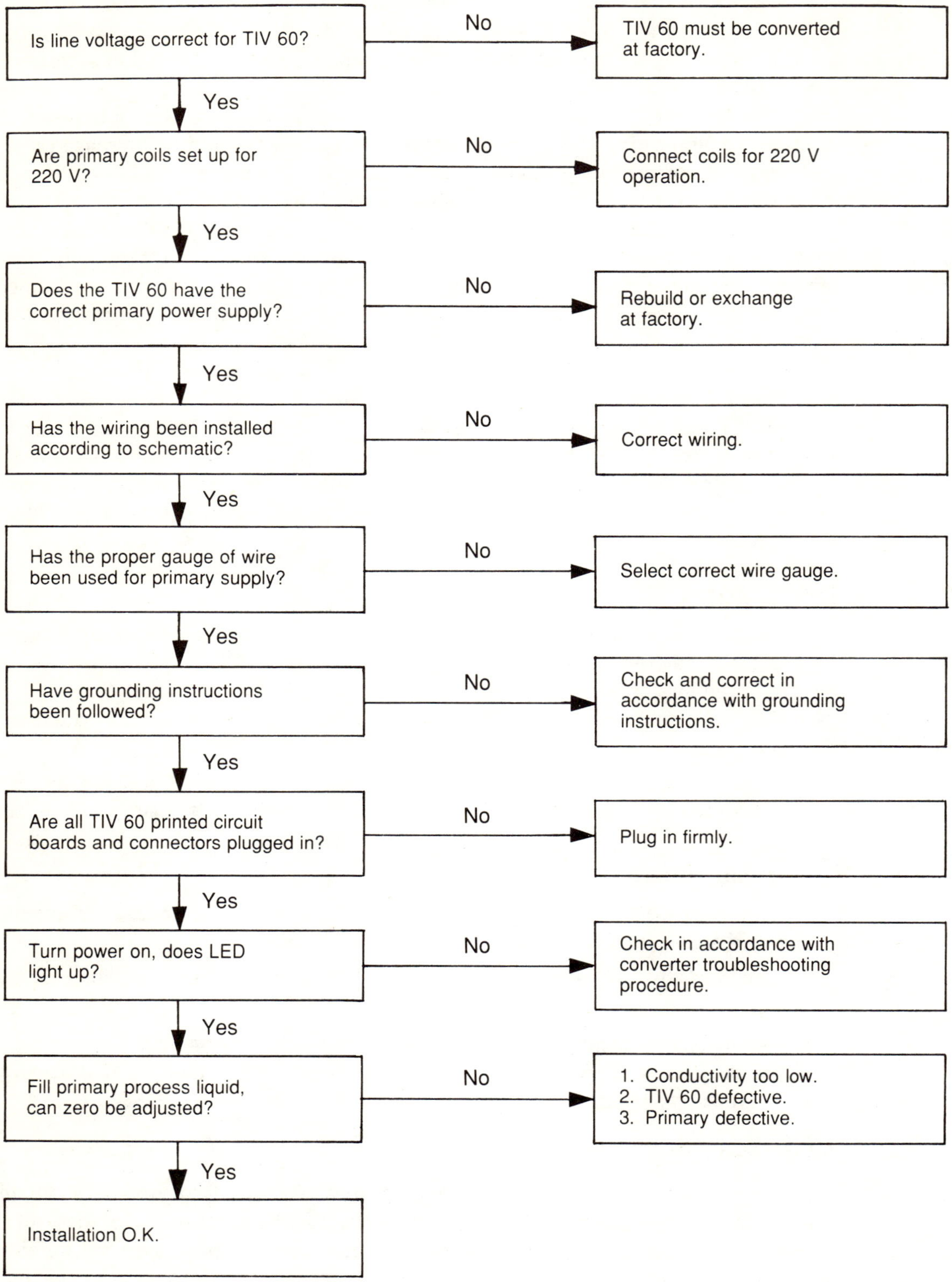

Adapted from Krohne-America, King of Prussia, Pa.

Figure 4-6 Flowchart of troubleshooting steps for installation check of magnetic flowmeter.

Some manufacturers provide special equipment for testing, troubleshooting, and recalibrating meters. Instructions and detailed diagrams are provided for disassembly and replacement of components. If a meter is considered to be not repairable in the field and the instructions indicate a warning against tampering with the sealed portion of the unit, no routine maintenance is recommended. Service or installation problems should be referred to the manufacturer's service or field office.

Electrode cleaning. Deposits and incrustations, including calcium carbonate, on the meter's electrodes can impair accuracy, unless the deposits have the same conductivity as the fluid. Such deposits can cause variable or high resistance between the electrodes, thereby introducing errors. Methods for cleaning the electrodes vary among the manufacturers. The following are examples of what may apply to most magnetic flowmeter electrodes:

- electrodes may be removed for inspection, cleaning, and replacement, without removing the meter from the line;
- conical electrodes are available that extend into the flow stream where the scouring effect of the liquid velocity is more likely to inhibit coating;
- continuous ultrasonic vibrations will prevent deposition of particulate and crystalline materials on the electrodes. This is not effective for grease deposits; and
- bypass piping to maintain flow may be included in the installation to allow for periodic inspection and cleaning of the meter's inner wall.

Recent advances in electronic signal conditioning have significantly reduced the effects of electrode fouling. Manufacturers now state that among the newer meters cleaning is not necessary, except under unusual circumstances.

Advantages and Disadvantages

The following are advantages of magnetic flowmetering devices:

- no obstruction to flow (unless meter spool size is reduced from pipe size and thereby causes head loss);
- minimum effective head loss, essentially that of the straight pipe replaced by the meter; magnetic flowmeters are highly suitable for applications where low head loss is essential;
- available over a wide range of sizes from 0.1 to 120 in. (2.5 mm to 3 m) in diameter;
- bi-directional, therefore, suitable for measuring reverse flows;
- linear output;
- variations in fluid density, viscosity, pressure, and temperature have little effect on performance; and
- unless very severe, upstream nonsymmetrical flow patterns and flow disturbances do not seriously affect the flow measurement.

The following are disadvantages of magnetic flowmetering devices:

- metered liquid must have an electrical conductivity of 20 μS/cm or greater (this is not a problem with finished drinking water);
- for smaller pipe sizes, the meters become relatively bulky and expensive;
- high accuracy is expensive, and each meter must be individually calibrated in a water test circuit; and

- the meter is sensitive to the geometry and electric properties of the flow tube and magnetic core, and to variations in the coil supply current.

TURBINE AND PROPELLER FLOWMETERS

Accuracy: ±0.5 to 2 percent of rate
Repeatability: ±0.02 percent
Rangeability: 10:1
Size range: $^3/_{16}$–24 in. (5–600 mm)
Head loss: High
Relative cost: High

In turbine and propeller flowmeters, flowing water strikes rotor blades, which rotate at a rate proportional to the flow velocity. The rotor (turbine wheel) of a turbine meter generally fills the cross section of the pipe and is mounted to spin freely between two central bearings supported in the pipe wall (Figure 4-7). The rotor (propeller) of a propeller meter (see Figure 4-7) is mounted on bearings at the downstream end of the pipe. The tapered propeller nose faces upstream into the flow and is mounted to spin freely in line with the pipe axis.

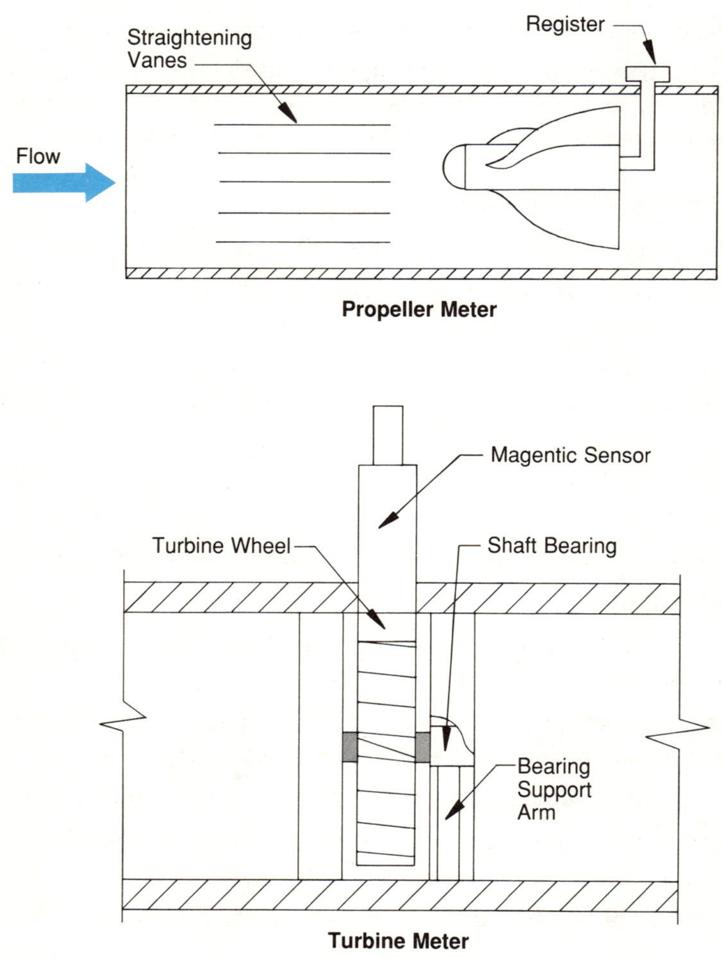

Figure 4-7 Schematic of turbine and propeller flowmeters.

Within given limits of flow rate and fluid viscosity, the rotor speed and volumetric flow rate maintain a linear relationship. A magnetic proximity sensor in the meter transduces the rotor velocity to an equivalent frequency signal. Hence, the rotation of the rotor blades causes a known number of cycles per unit volume. The meter coefficient is the calibration *K* factor, which is a known number of pulses for a given volume measured. The *K* factor is typically constant over a 10:1 flow range within a linearity tolerance of ±0.25. For identical meters, the *K* factor may vary as a result of manufacturing tolerances. Hence, each meter should be calibrated for its own specific *K* factor value.

According to meter manufacturers, turbine meter accuracy varies from ±0.5 percent to ±2 percent of flow. Among the manufacturers of propeller meters, there is agreement on accuracy of ±2 percent of actual flow. At low flow rates, accuracy is reduced to ±5 percent. The rangeability in larger meter sizes varies with good repeatability at 10:1, and can go up to 25:1.

The orientation and configuration of the meter blade profile are important to the application of the turbine meter. Straight-bladed meters may be less affected by variations in velocity, while helical-bladed meters are generally less affected by variations in viscosity.

Installation

Turbine and propeller flowmeters may be installed with a strainer to prevent solids from interfering with the rotor mechanism. Because these meters are affected by upstream configurations that cause swirls or velocity fluctuations, manufacturers frequently provide or recommend built-in straightening vanes upstream in the pipe. These are installed to minimize the effect of profile irregularities and to smooth flow entering the meter. Straight pipe lengths of at least five pipe diameters upstream and two pipe diameters downstream are recommended.

When selecting a meter for a specific application, care should be taken to ensure that the maximum flow rating will not be exceeded, except for short periods. Running over the maximum speed for extended periods will increase bearing wear and shift the meter's coefficient (*K*).

Location near a point of chemical injection should be avoided, and the electronic mechanisms that generate the pulses should be protected from electromagnetic influence.

Maintenance

A turbine or propeller flowmeter will demonstrate long life providing periodic mechanical maintenance is performed. Factory maintenance programs are available for meter testing, maintenance, and recalibration. Periodic inspection, calibration, and service should be performed at least once a year.

Propeller meters are generally manufactured for ease of disassembly and extraction of the metering unit from the pipe body. No special tools are required for maintenance, according to many manufacturers. When the metering unit is removed, a cover plate is commonly installed to continue line service. During service and disassembly, the complete metering assembly should be examined for wear and corrosion. Parts should be cleaned, and worn or damaged parts replaced. Troubleshooting procedures are provided by manufacturers, covering problems, causes, and recommended corrective actions. An excerpt from the troubleshooting procedures for a turbine meter is shown in Figure 4-8 (page 22).

Troubleshooting—Turbine Meter

INTRODUCTION

Turbine meter system malfunctions are usually restricted to two areas: electrical/electronic, or mechanical.

When a malfunction or an apparent malfunction occurs, the electrical and electronic systems should first be thoroughly checked in accordance with the manufacturer's recommended procedures, prior to checkout of the turbine meter. Only when the source of the malfunction cannot be found in the electrical or electronic systems should the turbine meter be inspected.

SYMPTOM: Fluid Delivery Greater Than Indicated on Totalizer

NOTE: Verify that the proper "K" factor value is entered in the electronic readout device.

1. **Possible Cause**—Foreign material collected on rotor or bearings. When foreign material collects on or in the bearings or if material wraps around the rotor (such as strands of PTFE tape), the angular velocity of the rotor will be reduced. This allows more fluid to pass through the meter than the pulse train indicates.

   Corrective Action—Remove the meter from the line and visually inspect internally. If foreign material is present, remove the material. If no foreign material is found, disassemble the meter in accordance with the instructions of Paragraph 4-3. Clean parts according to Paragraph 4-1. Reassemble and reinstall the meter in the line.

   If no foreign material is found, and cleaning does not eliminate the problem, check bearing wear as performed in the next paragraph.

2. **Possible Cause**—Excessive bearing wear. Excessive bearing wear will lower the angular velocity of the rotor and rotation may stop completely. Effect is the same as with foreign material.

   Corrective Action—Replace bearings or bushing assembly and journal. Recalibrate as required in accordance with Paragraph 4-2.

SYMPTOM: Fluid Delivery Less Than Indicated on Totalizer

NOTE: Verify that the proper "K" factor value is entered in the electronic readout device.

1. **Possible Cause**—Ground loop in electrical circuit.

   Corrective Action—See manufacturer's installation manual.

2. **Possible Cause**—Gasification of liquid in meter. ITT Barton recommends that a back pressure be applied to the system. This back pressure should be twice the magnitude of the net pressure loss through the meter plus twice the magnitude of the vapor pressure of the liquid being metered. If this back pressure is not maintained, gasification within the meter can occur, causing an over-spin of the rotor. This will indicate a greater than actual delivery and cause damage to the bearings.

   Corrective Action—Provide a back pressure at the meter by using accepted design means and practices.

3. **Possible Cause**—Entrained gas or bubbles in metered liquid. If the fluid has a significant amount of entrained gas or gas bubbles that are released in the reservoir, over-registration can occur. Turbine meters measure the actual volume of the liquid. If a unit volume is expanded by entrained gas or bubbles, this expansion will be registered. Gas bubbles may cavitate the rotor and produce effects similar to those caused by gasification of liquid in the meter (see previous paragraph).

   Corrective Action—Eliminate entrained gases or bubbles by use of an air eliminator or other acceptable method.

Adapted from *Turbine Meter Troubleshooting Procedures* (ITT Barton, City of Industry, Calif.)

Figure 4-8 Excerpt from turbine meter troubleshooting procedures.

Advantages and Disadvantages

The following are advantages of turbine and propeller flowmeters:

- the very large sizes have excellent short-term repeatability, and can be extremely accurate when calibrated periodically;
- output flow signal is directly proportional to a pulse train with high linearity over a broad range of flow rates, at least 10:1 for large meters;
- head loss is moderate, decreasing with larger sizes;
- meter size is the same as the diameter of pipe in which it is installed; and
- flow is not blocked if the meter seizes up.

The following are disadvantages of turbine and propeller flowmeters:

- systematic mechanical maintenance and lubrication are required;
- expensive, particularly for use in large-diameter pipe;
- wear or fouling of meter surfaces will gradually deteriorate the calibration, making periodic recalibration necessary to maintain high accuracy. Corrosive liquids, liquids of poor lubricating quality, and liquids with a high proportion of suspended solids will cause bearing problems;
- calibration factor is sensitive to changes in viscosity of the flowing liquid;
- sensitive to flow disturbances and swirls; and
- bearing friction is detrimental to performance of smaller meters.

SONIC FLOWMETERS

Accuracy: ±1 to ± 2.5 percent of rate
Repeatability: ±0.25 percent
Rangeability: 20:1
Size range: 0.125–120 in. (3.1 mm–3 m)
Head loss: None
Relative cost: High

In a sonic (or ultrasonic) flowmeter, a pair of transceivers (transmitter–receiver) are positioned diagonally across the meter body, as shown in Figure 4-9 (page 24). The transceivers transmit and receive an ultrasonic pulse in the direction of flow, followed by a return pulse against the direction of flow. In a flowing liquid, the speed of the pulse directed downstream is increased by the speed of the stream; when directed upstream, the speed of the pulse is slowed by the streamflow. The time difference between the two is a function of fluid velocity and, by computation, the rate of flow. The transit time difference and the allowance for pipe diameter are converted by a microprocessor circuit in the meter to a standard output signal for volume flow. This sonic meter, used to measure the flow of clean water, is known as a "time-of-flight," "transit-time," or "through-transmission" meter. The sonic meter accuracy is ± 1 percent of rate over a 10:1 range. Accuracy must be checked in the manufacturer's literature, since some manufacturers cite accuracy in terms of percent full scale.

Because sonic flowmeters use microprocessor circuits to generate and record flow, special digital features can be incorporated into the unit. Light-emitting diodes (LEDs) can indicate circuit operation or alarms, and provide readouts. The user can select output units, ranges, totalizing, and the type of signal transmission.

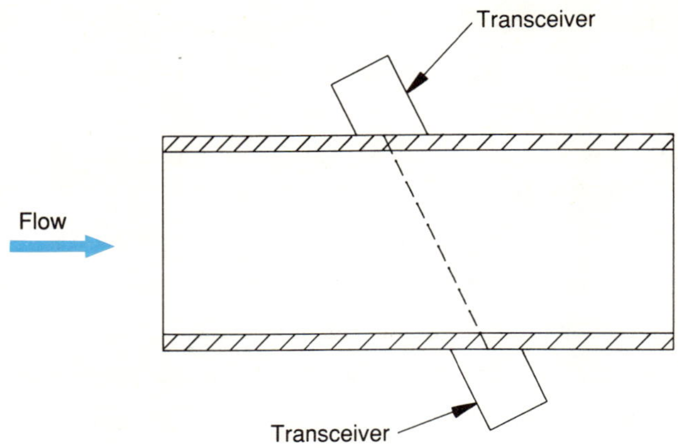

Figure 4-9 Schematic of an ultrasonic time-of-flight flowmeter.

The head loss of the meter is no more than that of an equivalent length of pipe. In cases of low flow rates, a less-than-line-size meter may be used, which will introduce some additional head loss.

Since there is no constriction, as in a Venturi or flow tube, there may be no flow conditioning. Therefore, any profile error caused by the piping configuration is carried through the meter and picked up by its velocity-sensing system. However, flow straighteners may alleviate the problem of turbulence. Some sonic meters use a single-path sensor system that requires more careful layout of upstream piping. To minimize profile sensitivity, some meters have multiple-path sensing to average out the fluid profile error, and thus increase metering accuracy.

Since the meter does not have a discharge coefficient that is defined by its hydraulic shape, all units, even of the same line size, must be calibrated.

The meter body is constructed of materials that have a long life. Because sonic flowmeters have been on the market for a relatively short time, the amount of performance data associated with field tests is not extensive. Consequently, the life cycles of the transceivers and related electronics over long periods of time have yet to be proven by experience.

One type of ultrasonic flowmeter is the Doppler, which requires only one transceiver to send pulses into the flow. This meter depends on particles of solids or entrained bubbles in the fluid to reflect energy pulses back to the transceiver. Consequently, this type of meter is not suitable for potable water.

Installation

A sonic meter requires approximately 10 pipe diameters of straight run upstream and 5 pipe diameters downstream for proper performance. This may differ with varying piping configurations and manufacturers. The meter should not be located near a point where there is a sudden pressure drop that might release minute quantities of gas from the liquid. The meter cannot function accurately when gas bubbles are present. Piping should be isolated from noise and vibration that might interfere with the sound propagation of the meter. The meter requires a power source and may use from 8 to 50 W of power.

The meter is available in two forms: a spool piece with integral transceivers or a transceiver assembly for clamp-on mounting outside an existing pipe.

Maintenance

Maintenance of sonic flowmeters is minimal. Sonic flowmeters should have a long operational life because of obstructionless flow, solid-state circuitry, and no moving parts.

Built-in electronic self-diagnostics generate circuit tests and display functions, and plug-in modular construction permits rapid replacement of defective parts.

Maintenance procedures are executed through the microprocessor, checking and adjusting the 4 and 20 mA output levels, and the level of the transmitted and received signal. A low signal level may indicate incorrect transducer installation, obscured sonic path (bubbles, solids), sedimentation, or cabling fault.

Advantages and Disadvantages

The following are advantages of sonic flowmeters:

- no obstruction to flow, thus, no head loss;
- not restricted to use with conductive liquids (as are magnetic flowmeters);
- clamp-mounted meters do not jeopardize pipe wall structure;
- clamp-mounted meters do not interrupt process flow during maintenance or replacement;
- no mechanical moving parts;
- linear output over a wide range (100:1);
- adaptable to a wide range of pipe diameters;
- low installation and operating costs; and
- bi-directional flow is allowable.

The following are disadvantages of sonic flowmeters:

- sensitive to change in fluid composition;
- high solids content or entrained bubbles distort and block propagation of sound waves;
- measures mean velocity across a diameter, which is not the same as the weighted mean velocity;
- sensitive to flow-velocity profile; accuracy can be impaired by changes in pipe wall roughness and by changes from laminar to turbulent flow;
- accuracy impaired by upstream and downstream flow disturbances, such as elbows and valves, which affect the velocity profile;
- positioning of the opposing transceivers is critical to ensure signal interception; and
- in clamp-mounted use, the presence of sound absorptive or scattering scale or coating on the inner walls of the pipe may prevent the meter from working (This is not true when transceivers are mounted through the wall on a spool piece.).

VORTEX FLOWMETERS

Accuracy: ±0.75 percent of rate
Repeatability: ±0.15 percent

Rangeability: 40:1
Size range: 1–10 in. (25–250 mm)
Head loss: Medium
Relative cost: High

In a vortex flowmeter, a nonstreamlined or bluff body, the vortex shedding element obstructs and splits the flow through the pipe forcing two streams around the barrier, and creating vortices downstream in the flow. These vortices are caused by the swirling of the fluid into the low-pressure area behind the body (see Figure 4-10). The vortices alternately rotate in opposite directions, with the spacing between them proportional to the fluid flow velocity. This also creates an oscillating pressure variation from side to side of the immersed vortex device.

Numerous methods are available for measuring the vortex-train frequency or the frequency of the pressure oscillations. In all cases, external electronics convert the frequency signal into a standard analog value that is proportional to the flow velocity or a pulse train suitable for input to a totalizer.

Accuracy of vortex flowmeters is ±0.75 percent of rate, with a repeatability of better than ±0.15 percent. Flow rate is approximately 15 ft/s (5 m/s), with a range of 40:1. The flow range is a function of pipe size, being dependent on the Reynolds number and on cavitation in the pipe.

Head loss is somewhat higher than that of an unobstructed pipe, because of the presence of the vortex element. The added loss is equivalent to about 4 psi (28 kPa) at maximum flow.

The frequency/flow characteristic in the pipe is a function only of the shape of the body. Consequently, a generic coefficient can be used for all meters having the same body profile, regardless of pipe size.

Pipe sizes used with vortex flowmeters range from 1 to 10 in. (25 to 250 mm), and larger sizes can be used. The maximum size of pipe that can be used depends on the pulse frequency per unit volume. This limit exists because, in larger pipes, the full-scale frequency may be too low to enable the signal-conditioning electronics to make acceptably accurate measurements.

The vortex meter has no moving parts, and has only the spool piece and body exposed to the fluid. It is typically constructed of stainless steel, resulting in relatively low maintenance requirements.

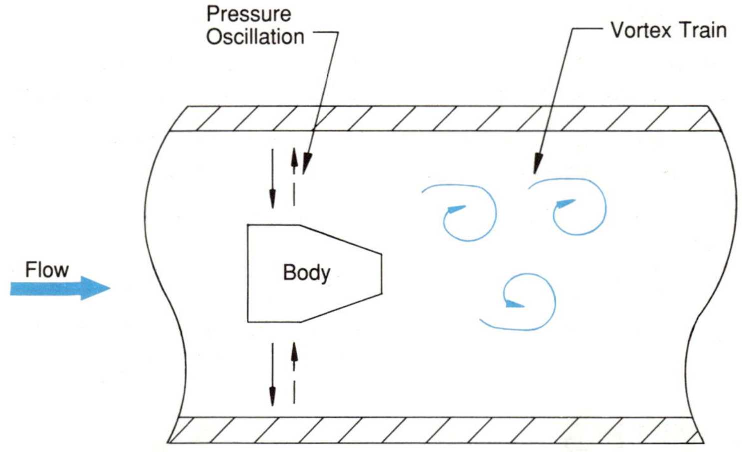

Figure 4-10 Schematic of a vortex flowmeter.

Installation

The vortex meter requires a fully developed flow profile, which means that upstream disturbances must be minimized. An upstream straight pipe length of at least 10 diameters is desirable, and flow straighteners may be needed where severe disturbances are present. A length of five diameters of straight pipe downstream should also be used, to minimize the effects of disturbances on the vortex train. Severe vibration of the pipe caused by "noisy" pumps and valves and the continued presence of bubble streams can affect meter accuracy by introducing false signals into the sensing elements. These problems must be considered before installation of the meter.

Maintenance

The meter is not expected to fail if properly installed and operated under normal conditions. If failure occurs, flowcharts of troubleshooting instructions and related diagrams are provided by manufacturers to isolate and remedy the problem. An example of a flowchart for troubleshooting is shown in Figure 4-11 (page 28).

Parts are usually removable, according to detailed procedures found in the instruction manuals, and replaceable in the field. These include the vortex shedding element, the sensor assembly, and output signal electronics. The electronics can be replaced without interrupting the pipe flow.

Because it has no moving parts, there is virtually no maintenance associated with the installed fixed assembly of meter spool and vortex shedder. This assumes that installation is meticulously executed according to manufacturer instructions regarding orientation, alignment, piping connections, ambient conditions, power connections, wiring, and flow range applications.

Detailed instructions for maintenance usually refer to electronic adjustments, such as zero, span, noise balance, and minimum measurable velocity. However, all such adjustments, while explained in the meter manual, are set in the factory and should not be altered after proper installation.

Advantages and Disadvantages

The following are advantages of vortex flowmeters:

- low head loss;
- no moving parts;
- long-lived construction;
- high turndown ratio (the ratio of the maximum design flow rate to the minimum design flow rate); and
- simplicity in design and installation.

The following are disadvantages of vortex flowmeters:

- sensitive to flow-profile distortion;
- affected by pipe vibration and bubble streams;
- not bi-directional; and
- limited range of pipe size.

AVERAGING PITOT FLOWMETERS

Accuracy: $\pm 1/2$ to ± 5 percent of full scale
Repeatability: 0.5 percent

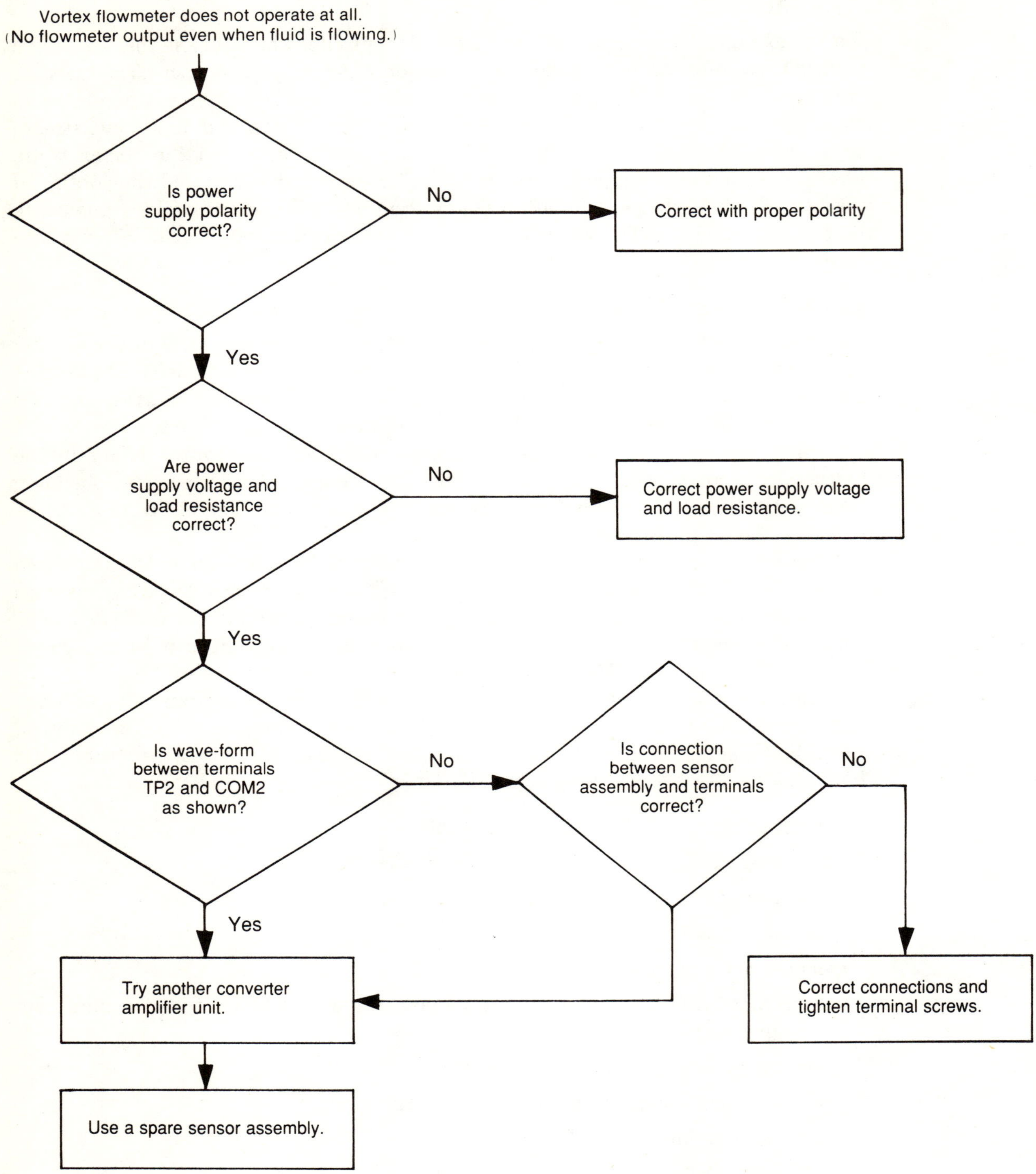

The vortex flowmeter seldom fails under normal operating conditions. Improper installation of the instrument or pipeline may cause troubles. If it does not operate properly, carefully check, isolate, and remedy troubles as per appropriate troubleshooting flowcharts. If it is difficult to remedy any troubles, contact your nearest service center.

Adapted from Yokogawa Corporation of America, Shenandoah, Ga.

Figure 4-11 Troubleshooting flowchart for vortex flowmeters.

Rangeability: 4:1
Size range: $^1/_2$–96 in. (13 mm–2.4 m)
Head loss: Low
Relative cost: Low

The averaging Pitot flowmeter consists of an insertion tube, or probe, that is placed through the pipe cross section (Figure 4-12). Multiple ports face upstream into the flow to provide sampled pressures at selected points along the pipe diameter. The multiple pressures are sensed as an interpolated average by the internal tube to provide an averaged pressure over the pipe cross section. A port or ports facing downstream register static pressure. The device produces a differential pressure between the averaged velocity head ports and the static head port(s). As with the Venturi flowmeter, the fluid velocity is proportional to the square root of the differential between the resultant upstream pressure and the static pressure.

Accuracy for a Pitot meter is $\pm^1/_2$ percent to ± 5 percent of full scale. The meter's range is greatly limited by the sensitivity of the differential-pressure sensor and the results obtainable for low differential. The meter, by nature, is velocity-profile sensitive. Placing it in different planes at the same point in a pipe may result in different readings. Laboratory calibration is not valid for field conditions unless the pipe configuration is duplicated.

According to manufacturers, the shape of the insertion tube generally will cause most foreign material to flow around the probe rather than accumulate on it. Ordinarily, material that does affect the probe does not significantly affect meter performance unless, in extreme cases, the ports are completely obstructed or buildup changes the outside shape.

Installation

The insertion tube must be installed into the pipe according to the manufacturer's instructions. Deviation from perpendicular to the axis of the pipe in any direction will

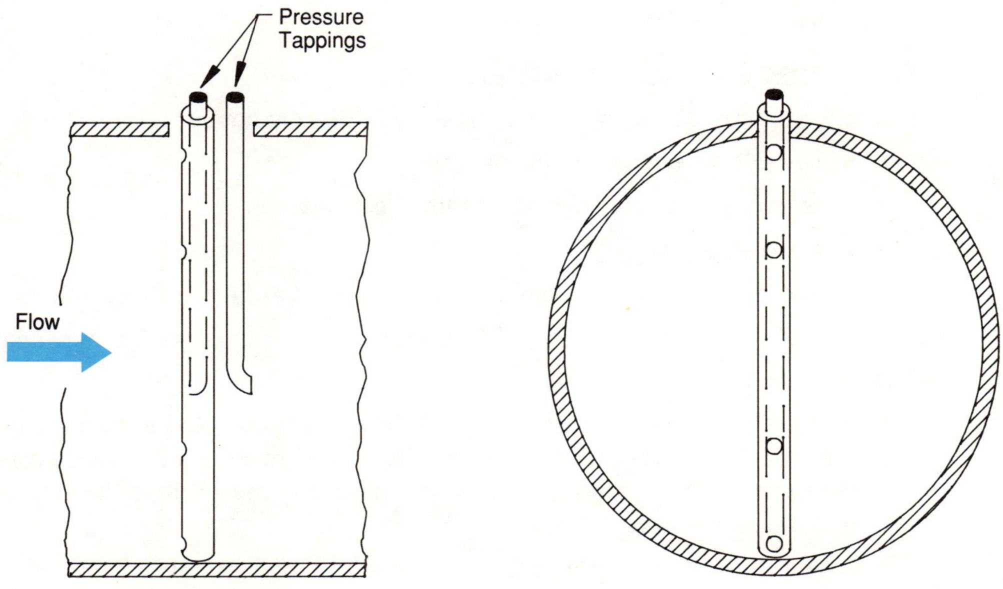

Figure 4-12 Schematic of an averaging Pitot flowmeter insertion tube.

affect the sensed pressures. Misalignment of the tube from the diameter beyond 3° in the pipe cross-sectional plane or 5° upstream or downstream out of the crosssectional plane will cause significant error. Rotation of the tube beyond 3° from strict upstream/downstream will also cause errors in the sensed pressure.

It is also important to install the correct size insertion tube in the correct size pipe. The location of the ports on the insertion tube is based on the pipe's inside diameter and wall thickness. Installed in the proper size line with the proper fittings, the tube's sensing ports will be at their correct locations, and the meter will respond with the designed accuracy. If the tube is inserted in an incorrect line size, it can be expected to provide a repeatable signal, but must be recalibrated to yield accurate flow measurements.

The location of tube ports is based on a fully developed turbulent flow, with a velocity profile that is consistent across the pipe in all directions. The averaging of pressures will not be correct in an inconsistent flow profile, and errors in flow measurement will result. Upstream devices, such as valves, elbows, and pipe diameter changes, influence the flow profile. Therefore, as with other meters, sufficient lengths of pipe must be provided upstream of the insertion tube to allow the turbulent flow profile to develop. Flow straighteners may be used to reduce the necessary length of straight pipe upstream. Tables of recommended upstream and downstream straight pipe lengths are provided by manufacturers. Upstream lengths vary from 7 to 24 pipe diameters without straighteners, and 3 to 9 pipe diameters with straighteners. Downstream straight pipe lengths vary from three to four pipe diameters.

Maintenance

The probe should be removed and cleaned periodically. Probe removal does not require shutting down the system. Sensing ports and internal passages can be cleaned using external pressure without being removed. Manufacturer's recommendations should, of course, be followed in cleaning the probe. The sensor design will handle most flow conditions without clogging. Nevertheless, if the fluid is contaminated, periodic purging of the internal passages may be necessary. Designs can be provided to facilitate purging.

Advantages and Disadvantages

The following are advantages of averaging Pitot flowmeters:

- removable without shutting down system;
- nonconstricting design produces low head losses;
- meter cost is low;
- easily installed at any time by making a wet tap in the pipeline; and
- materials of construction and the nature of the parts provide for long life.

The following are disadvantages of averaging Pitot flowmeters:

- any leaks in the instrument lines or connections will significantly affect the meter accuracy, because the flow measurement depends on the comparison of two pressures generated by the flow past the device, and the pressure change may be small; and
- potential for calibration error exists due to misalignment of the insertion tube.

Bibliography

Automation and Instrumentation. AWWA Manual M2. AWWA, Denver, Colo. (1983).

BEAN, H.S. Fluid Meters: Their Theory and Application. Rept. of ASME Res. Comm. Fluid Meters, Amer. Soc. Mech. Engrs., New York (6th ed., 1971).

CASCETTA, F. & VIGO, P. *Flowmeters, A Comprehensive Survey and Guide to Selection*. Instrument Society of America, Research Triangle Park, N.C. (1988).

CHEREMISINOFF, N.P. & CHEREMISINOFF, P.N. *Instrumentation for Process Flow Engineering*. Technomic Publishing Company, Inc., Lancaster, Pa. (1987).

HAYWARD, A.T.J. *Flowmeters, A Basic Guide and Source Book for Users*. John Wiley & Sons, New York (1979).

JOHNSON, C.D. *Process Control Instrumentation Technology*. John Wiley & Sons, New York (1988).

LIPTAK, B.G. *Instrument Engineers' Handbook*. Vol. I. *Process Measurement*. Chilton Book Company, Philadelphia, Pa. (1982).

———. *Instrument Engineers' Handbook*. Vol. II *Process Control*. Chilton Book Company, Philadelphia, Pa. (1985).

MILLER, R.W. *Flow Measurement Engineering Handbook*. McGraw-Hill Book Company, New York (1983).

ZIMMERMAN, R. & DEERY, D. *Handbook for Turbine Flowmeter Systems*. Flow Technology, Inc., Phoenix, Ariz. (1977).

Index